H. Joshua Phillips

**Fuels; Solid, Liquid and Gaseous**

Their Analysis and Valuation

H. Joshua Phillips

**Fuels; Solid, Liquid and Gaseous**
*Their Analysis and Valuation*

ISBN/EAN: 9783337139834

Printed in Europe, USA, Canada, Australia, Japan

Cover: Foto ©ninafisch / pixelio.de

More available books at **www.hansebooks.com**

# FUELS

## SOLID, LIQUID, AND GASEOUS

## *THEIR ANALYSIS AND VALUATION*

### FOR THE USE OF CHEMISTS AND ENGINEERS

BY

## H. JOSHUA PHILLIPS, F.C.S.

ANALYTICAL AND CONSULTING CHEMIST TO THE GREAT EASTERN RAILWAY
FORMERLY CHIEF ASSISTANT
IN THE LABORATORY OF THE GREAT WESTERN RAILWAY

Capio Lumen

LONDON

## CROSBY LOCKWOOD AND SON

7, STATIONERS' HALL COURT, LUDGATE HILL

1891

# PREFACE.

THE great progress made in recent years in the adaptation of various kinds of fuel to special purposes, is sufficient, it is believed, to warrant the issue of a small work dealing with methods of analysis and valuation of fuel, and summarising some of the results already ascertained.

The object of this little book is to comprise within small limits that which would otherwise have to be gleaned from widely scattered sources and larger works; and the Author's endeavour has been to make it a practical and theoretical help both to the chemist and the engineer, by whom the analytical tables and practical results set forth in the volume will no doubt be found of service.

He has to acknowledge his obligations to the various analytical works and articles in English and foreign journals, and the proceedings of societies, &c., at home and abroad, which have been consulted.

Should the work meet with the approbation of those who are interested in this important subject, the Author can see his way to its extension and improvement, and the increase of its usefulness, if it should have the honour of running to a second edition.

TAVISTOCK SQUARE,
    *October,* 1890.

# CONTENTS.

## METHODS OF ANALYSIS.

## ANALYSES OF VARIOUS GASEOUS FUELS.

# THE
# ANALYSIS AND VALUATION OF FUELS.

———•———

AT a time when fuel has become such an expensive commodity, the consideration of its quality, heating power, and economic application becomes a matter of great importance to engineers and steam users in general.

In determining the value of a solid fuel such as coal, coke, or patent fuel, it is necessary to estimate the amount of moisture, volatile matter, coke and its quality, sulphur, the amount of ash left on the incineration of sample, together with the calorific and evaporating power ascertained—(1) Theoretically, from a knowledge of the percentage of hydrogen, carbon, and oxygen found to be present by an elementary organic analysis of the substance; or (2) Practically, by igniting a known weight in a calorimeter in oxygen and ascertaining the amount of ice that has been melted, or observ-

B

ing the increase of temperature of a known weight of water; or (3) By ascertaining the amount of lead reduced from its oxide by a known weight of the sample. This method, however, is not often adopted for calorific power, the amount of lead reduced being a measure of the reducing power of a sample, rather than the amount of heat that it is capable of evolving.

The calorific and evaporating power of fuels estimated by the foregoing methods must only be taken relatively. The actual highest practical value of any fuel would depend very materially upon the kind of furnace used, and the regulation of an appropriate influx of air, so as to insure complete combustion. The heat carried up the flues by the products of combustion should, if possible, be reduced to a minimum, so as to obtain the most economic results.

Liquid fuels are now coming to the front, and judging from recent experiments made on locomotives in South Russia, where petroleum is plentiful and cheap, and coal comparatively dear, they seem to have the advantage over coal. The specific gravity, flashing point, and the calorific power obtained practically by slightly modifying the process as used in the case of coals, or by calculating it theoretically from the percentage of carbon and hydrogen found to be present, would be valuable data for guidance as to their selection. Sulphur, generally speaking, is present in so small a quantity as to be not worth consideration.

With regard to gaseous fuels, the principal heat-giving ingredients usually present are hydrogen, gaseous hydrocarbons (marsh gas, olefiant gas, &c.), and carbonic oxide. The determination of these, and the calculation of their respective calorific values on combustion, would convey an estimate of their efficiency for heating purposes.

### Estimation of Moisture and Ash (Solid Fuels).

Weigh out 3 grams of the fairly averaged powdered sample in a shallow platinum dish, and dry in an air bath regulated to 105° C. for one hour, allow to cool in a good desiccator, and weigh; the loss is taken as moisture.

Coal and cokes should not be allowed to dry for more than one hour, otherwise the result obtained for moisture will be too low, owing to the oxidation of the pyrites present by the air.

The dish with the dried sample is now cautiously heated to redness in a muffle until it is seen that all carbonaceous matter has been eliminated, and then allowed to cool in the desiccator and weighed. It is once more heated in a muffle for about five minutes, and again cooled and weighed.   If the weight is unaltered from that of the last weighing, the increase in weight over the dish is the amount of ash left by 3 grams of sample, which can easily be expressed centesimally.   The amount of ash left by different samples is very variable, some coals showing only 7 per cent., while others as much as 20 per cent.

### Estimation of Coke and Volatile Matter.

1 gram of the sample is carefully weighed into a small porcelain crucible provided with a cover, and cautiously heated to redness by a large Bunsen burner for two minutes, and then heated for an additional two minutes at a higher temperature over a gas blow-pipe; it is then allowed to cool in a desiccator and weighed.   The loss $=$ moisture $+$ volatile matter, and the residue $=$ coke $(+$ ash$)$.

The coke is tested by means of a penknife, to ascertain if it is friable or compact.   A little of it is placed on platinum foil and ignited, and noted as to whether it burns freely or not.

Mr. G. E. Davis makes an interesting classification of coals according to the amount of coke that they are capable of producing :—

(1) Splint coal, burning with a long flame, and

yielding from 50 to 60 per cent. of powdery or slightly caked coke.

(2) Gas coal, or coal of a bituminous nature, burning with a long flame, and yielding from 60 to 70 per cent. of fused but deeply seamed coke.

(3) Smithy coal, or true bituminous coal, burning with a long flame, and yielding 68 to 75 per cent. of fused compact coke.

(4) Caking coal, burning with a short flame, leaving from 75 to 82 per cent. of fused compact coke.

(5) Anthracite or smokeless steam coal.

## Estimation of Sulphur.

Sulphur is a very deleterious constituent of coal and coke, both for boilers and metallurgical purposes. Very small quantities finding its way into "pig" iron will render the latter unfit for steel-making, while the sulphurous vapour formed by its combustion in boilers corrodes fireboxes, boiler-tubes, &c.

The disagreeable odour in our underground railway tunnels is largely attributable to compounds of sulphur (sulphuretted hydrogen, bisulphide of carbon, &c.), mainly brought about by the distillation of fresh portion of coal thrown on

to the red-hot coke in the fireboxes of the locomotives.

Sulphur exists in two forms in coals and cokes,
being present as iron pyrites (Fe $S_2$), and techni-
cally termed " brasses," and sulphate of lime
(Ca $SO_4$).

The sulphur present as iron pyrites alone appears
to affect the economic application of the fuel.

To determine the total sulphur present, weigh
out 2 grams of the sample and mix thoroughly
with 5 grams of pure powdered nitrate of potash,
and add this mixture in small portions at a time to
8 grams of pure anhydrous sodic carbonate kept in
a steady state of fusion in a capacious silver crucible by means of the oxidising flame of a Bunsen
burner. The crucible should be tilted, and the
flame kept as far from its mouth as possible, to
prevent access of sulphur compounds from the gas.
When the mixture has been fusing for about ten
minutes, after the addition of the last portion of
the sample, it is acidulated with dilute hydrochloric
acid, evaporated to dryness, heated on sand bath
to about 130° C. to render silica insoluble, mois-
tened with 10 cc. of strong HCL, diluted with
distilled water to 100 cc., filtered and washed, and
the filtrate diluted to about 500 cc., heated nearly
to boiling. A few cc. of a saturated solution of
Baric chloride is now added, well stirred and

allowed to stand in a warm place if possible, for about twelve hours. The supernatant liquid is now syphoned off and the precipitated Baric sulphate ($Ba\ SO_4$) carefully filtered on to a No. 2 Swedish filter paper, and washed till free from soluble matter with hot water. The filter paper is then spread out on a watch-glass, and put to dry in a water oven; in the meantime a small porcelain crucible is carefully weighed. When the filter is dry the $Ba\ SO_4$ is brushed into it, the filter burnt separately and its ash added to the main portion; the crucible and its contents are then ignited at a dull red heat in the oxidising flame, allowed to cool, and weighed. Increase in weight $= Ba\ SO_4 +$ ash, subtract ash, then $\dfrac{Ba\ SO_4 \times \cdot 1373 \times 100}{2} =$ percentage of total sulphur.

The sulphur present as sulphate of lime is determined by boiling 5 grams of sample with a strong solution of sodic carbonate; the sulphur is by this means converted into soluble sulphate of soda; dilute and filter, and estimate sulphur as above. On subtracting the sulphur thus found from the total, the amount of sulphur present as pyrites is obtained.

## Estimation of Nitrogen.

Owing to the small quantity of nitrogen usually present in fuels, it is best determined volumetrically. The following is the process devised by Dumas. A combustion tube is selected about 120 cm. long, and sealed at one end like a test

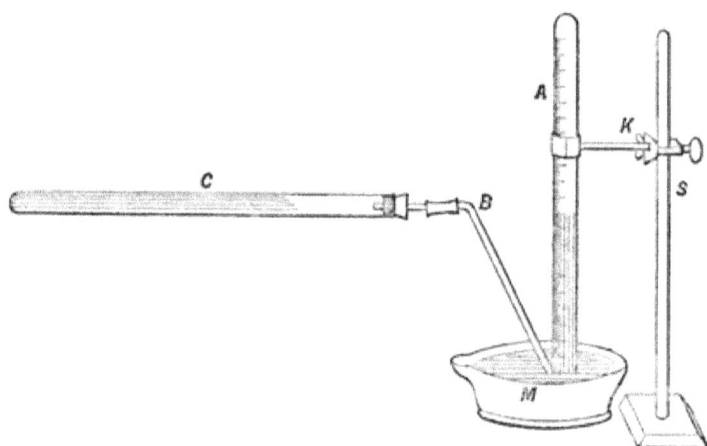

Fig. 1.

tube, cleaned and dried. A layer of pure bicarbonate of soda 15 cm. long is first introduced, then a layer of copper oxide 20 cm. long, after this an intimate mixture of 1·5 grams of the sample with oxide occupying 30 cm., then 30 cm. of coarse copper oxide and 20 cm. of small copper turnings; the tube is now connected by means of a good fitting cork with the bent delivery tube B, Fig. 1, and placed in a combustion furnace. The further end of the tube containing the bicarbonate is then gradually heated to the extent of 6 cm. Carbonic acid gas is evolved and sweeps the air out of the tube ; when the gas has been coming off for a few minutes the end of the delivery tube is dipped under mercury contained in the trough M, and the issuing gas tested for air by inverting a test tube filled with a strong solution of potassic hydrate over it. If the gas as it comes in is completely absorbed all air has been eliminated ; if not, continue the heating until the desired point is attained. The graduated tube A, which is filled with $\frac{2}{3}$ mercury and $\frac{1}{3}$ with a strong solution of potassic hydrate, is then inverted over the end of the delivery tube and held in position by the clamp K. The combustion is now proceeded with. The fore part of the tube containing the copper is first cautiously heated to redness, and the heat gradually extended to the further end until the point where the sample ends. When no more gas comes off from the sample, the other half of the bicarbonate is heated ; the second crop of $CO_2$ thus produced drives any nitrogen still left in the tube into A. When the volume of gas in the tube A is no longer diminished by the absorption of any $CO_2$ that may be still present, even on shaking, the tube is then transferred by means of a small dish filled with mercury to a deep vessel containing water. The mercury will then be displaced by water; the tube is pushed into the water until the levels of the liquid are coincident and the volume of nitrogen noted, together with the temperature and barometric pressure. The volume observed is reduced to 0° C. and 760 mm. pressure; and inasmuch as the gas is measured over water, allowance must be made for the pressure reduced by the tension of aqueous vapour at the temperature observed.

The following is an example of an actual determination of nitrogen in a sample of South Stafford coal:—

Volume of nitrogen observed . 25 cc.
Temperature     .     .     .     . 15° C.
Barometric pressure .     .     . 758 mm.

The tension of aqueous vapour at 15° C. is equivalent to 12 677 mm. of mercury.

Taking 1 cc. of nitrogen at 0° C. and 760 mm. pressure as weighing ·0012544 gram, then the percentage of nitrogen by weight in the sample is—

$$\frac{25 \times 273 \times (758 - 12\cdot677) \times \cdot0012544 \times 100}{(273 + 15) \times 760 \times 1\cdot5} =$$

1·943 per cent. nitrogen.

## Estimation of Carbon and Hydrogen.

The principle of the method adopted for the determination of carbon and hydrogen depends upon
the fact that when a fuel is burnt in excess of air
or oxygen, or any oxidising substance, the carbon
is oxidised into carbonic acid gas ($CO_2$), while the
hydrogen is converted into water ($OH_2$), evolved as
steam.  A known weight of the sample being
taken, it is ignited with chromate of lead or copper
oxide and oxygen or air; the products of combustion $CO_2$ and $OH_2$ are absorbed by appropriate
reagents and weighed separately, from which the
amounts of carbon and hydrogen present are deduced by a simple calculation.

Select a wrought-iron tube 20—22 mm. in
diameter and 1.5 cm. long; oxidise inner surface
of tube by heating it to redness in a combustion
furnace and passing a current of steam through; a
layer of recently ignited coarse oxide of copper,
about 20 cm. long, is pushed into the middle of
the tube and kept in position by plugs of copper
gauze placed one each side; a sheet iron boat, about
30 cm. long, is nearly filled with recently fused
and powdered chromate of lead and introduced into
the tube, which is placed in a combustion furnace
and heated below the fusing point of the chromate,
a current of dry air being passed through, to rid

the tube of any moisture. The gas is then put out, and the tube plugged and cooled. The boat is then taken out, and from ·3 to ·5 gram of the powdered sample free from water quickly and thoroughly mixed with the chromate, and replaced in the tube; a similar boat filled with recently reduced metallic copper is introduced at the other end of the tube, and the whole placed into the

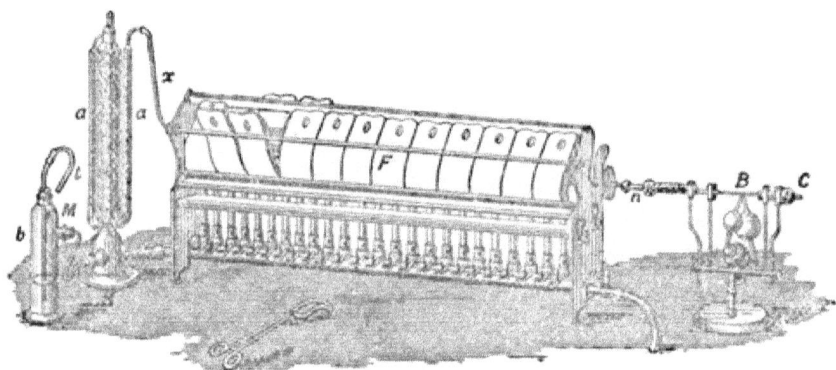

Fig. 2.

furnace and coupled up to the desiccating and absorption apparatus. Fig. 2 shows the apparatus ready for a combustion.

*b* is a cylinder containing a strong solution of potassic hydrate, which absorbs the great bulk of $CO_2$ present in the air used for the combustion ; it is coupled up to a gas holder by means of the tube *t*, and to the cylinder *k*, which is filled with fragments of soda lime, which complete the exhaustion of

$CO_2$. U tubes *a a* are filled with dry granulated
calcium chloride which completely absorbs atmospheric moisture, they are connected to the combustion tube by means of the pipe *x*; the bulb tube
*n* contains dry calcic chloride which serves to
absorb the water brought about by the combustion
of the hydrogen in the sample. The bulbs B contain a strong solution of potassic hydrate, which
absorbs the $CO_2$ produced by the combustion of
the carbon. *c* is a small tube containing calcium
chloride, which serves to retain any moisture
carried from B by the issuing air.

Before proceeding with the combustion, *n* and
B are accurately weighed separately and attached
to tube as in sketch; all joints having been
insured air tight, the gas is turned on at the end
of the tube containing the reduced copper, and a
gentle current of air made to pass through the
apparatus, when it is at a dull red heat; the gas
is then gradually turned on until the boat containing the sample is reached. Care is here required,
and the heat is gradually raised to full redness,
when the chromate will fuse, and the sample is
soon completely oxidised. After it has been
insured that the combustion is complete, the calcium chloride tube *n* and potash bulbs B are
detached and weighed. The increase in weight of
the Ca $Cl_2$ tube multiplied by ·1111 gives the amount

of hydrogen in quantity of sample taken, and the increase in weight of the bulbs B multiplied by ·27273 gives the amount of carbon present in the quantity of sample taken.

In estimating the carbon and hydrogen in non-volatile liquid fuels the two boats are filled with copper oxide and the weighed liquid absorbed in one of the boats, and the combustion proceeded with as above ; or if it be a volatile hydrocarbon it is weighed into a small thin tube, with a loose stopper and dropped into the boat, due care being taken that the combustion is not hurried, otherwise vapour will escape unoxidised.

The following is an example of the results obtained for carbon and hydrogen in a Scotch bituminous coal, when ·5 gm. was taken :—

Weight of Ca Cl$_2$ tube before combustion    30·4562

” ” after ” . 30·6884

Water absorbed    ·2322

·2322 × ·1111 × 2 × 100 = 5·159 per cent. hydrogen.

Weight of potash bulbs before combustion . . . . . . . 52·0318

Weight of potash bulbs after combustion    53·4296

CO$_2$ . . 1·3978

$$1\cdot3978 \times \cdot27273 \times 2 \times 100 = 76\cdot244 \text{ per cent.}$$
carbon.

## Estimation of Oxygen.

There is no ready method for the direct determination of oxygen in complex organic compounds. It is as a rule estimated by difference;
having a knowledge of the percentage of all other
ingredients present in the sample, add these up
and subtract from 100, and the remainder may be
taken as oxygen.

## The Specific Gravity of Coals, &c.

It is often desirable to know the amount of space
that a given weight of coal will occupy, and the
determination of the specific gravity thus becomes
necessary.

To obtain this a small flask provided with a
thermometer stopper, and holding a definite weight
of water at a known temperature—usually 60° F.,
which is previously accurately ascertained — is
taken, and 2 to 3 grams of the sample weighed
into it; water is then added, and the coal allowed
to soak in it for some time, so as to eliminate air
from the pores. The bottle is then filled with
water at the standard temperature and again
weighed.

The specific gravity is obtained as follows :—

Let $W$ = weight of sample in air;

$R$ = weight of flask + water;

$R_1$ = weight of flask + water + sample.

Then,

$$\text{Specific gravity} = \frac{W + R - R_1}{W}$$

The weight of a cubic foot of the sample in pounds is obtained by—

log. specific gravity + 1·79588 = log. weight of cubic foot.

The number of cubic feet in a ton

= 1·55437 — log. specific gravity = log. cubic feet.

It is very important in determining the specific gravity of coals and cokes to insure that all air has been driven out of the sample by the water before diluting and weighing. An example may be given of the error that would be involved if this precaution were not taken. Mr. Crookes, F.R.S., obtained the following: 2·76 grams of coal gave the specific gravity 1·309 at 64° F., immediately after filling the flask with water; after soaking twelve hours the specific gravity had increased to 1·328 for the same temperature. So that the latter determination would make a cubic foot of this coal weigh 82·76 lbs., and the former only 81·58, or 1·18 lbs. less.

## Specific Gravity of Liquid Fuels.

The specific gravity of liquid fuels can in the majority of cases be determined at 15° C. by the hydrometer in the usual manner. It sometimes happens, however, that liquid fuels are too thick to obtain an accurate result by this means, and it has to be determined by the specific gravity bottle ; or if too thick for this, by placing a drop in a cylinder of alcohol at 15° C., and adding water until it remains stationary in any part of the fluid in which it is placed, a glass rod being used for the purpose ; the specific gravity of the fluid is then taken with the hydrometer, and the specific gravity of the sample is thus attained.

## The Flashing Point.

The temperature at which the vapour coming off from liquid fuels ignites when mingled with air, on the application of a flame, is of some importance. The lower the temperature at which the vapour is capable of igniting, the more care, of course, will be required with its handling, storage, transport, &c.

A ready mode of determining the flash point of a sample is to pour some of the liquid into a beaker (2″ × 2″) to within about half-an-inch from the top; then cover with a disc of asbestos, through which a thermometer passes to within a quarter of an inch from the bottom of the beaker. The beaker, &c., is now put into a sand bath, and surrounded with sand to the level of the liquid; a small flame is then applied under the bath, and the temperature allowed to rise about 2° per minute. After each rise of 1°, the asbestos disc is turned to one side, and a small flame is quickly put into the vapour. The temperature at which it ignites is taken as the flash point.

A more definite method of testing the flash points of light oils is obtained by Sir Frederick Abel's standard flash-point apparatus, supplied by Townson and Mercer, London.

### Determination of the Calorific Value of Solid and Liquid Fuels by Thompson's Calorimeter.

In this method, which is now most extensively used in estimating the heating power of fuels, a known weight of sample is ignited with an oxygen mixture in a copper cylinder, in a known weight of water, the temperature of which is first accurately observed. From the increase of temperature of the water, due to the combustion of sample, the comparative heating and evaporating power can be deduced.

The French unit of heat is the amount of heat necessary to raise the temperature of 1 gram of water through 1° C., or more correctly from 0° to

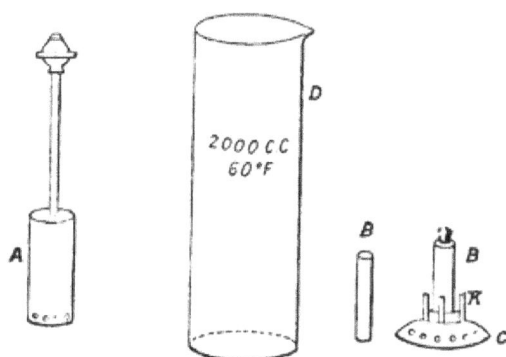

Fig. 3.

1° C. The calories as here expressed are the number of grams, lbs., or any unit weights of water raised 1° C. by the combustion of 1 gram or 1 lb., &c., of the sample. These can be converted into British Thermal units: viz., lbs. of water raised 1° F. by 1 lb. of sample, by multiplying by $\frac{9}{5}$.

For coals and cokes, 2 grams of the finely powdered and dried sample is thoroughly mixed with 26 grams of a finely powdered and dry mixture of

chlorate of potash 3 parts, and nitrate of potash
1 part, on a sheet of glazed paper. By means of
a flexible steel spatula, the mixture is introduced
in small quantities at a time into the copper
cylinder B, Fig. 3. Each addition being pressed
with about the same pressure with the rounded
end of a test-tube, so that a fairly uniform combustion can be relied upon. When all has been
brushed in, a short length of fuse (prepared by
soaking lamp cotton in a strong solution of nitre
and drying) is pushed into the mixture, and about
half an inch allowed to protrude. It is fixed into the
brass stand C. The glass cylinder D, having been
charged with 2000 cc. of water, the condenser A is
fixed over B into C, held firmly by the clutches K.
The whole is then placed into D, and moved up
and down in the water until the temperature of the
latter is fixed. The temperature is recorded by a
very delicate thermometer graduated to $\frac{1}{20}$ of a
degree centigrade.

The temperature of the room is usually higher
than the water, so that a little hot water or ice (as
the case may demand) must be added, until the
differences are about as follow :—

| Room at °C | 27 | 23 | 20 | 16 | 13 | 10 | 6 |
|---|---|---|---|---|---|---|---|
| Water should be | 21 | 18 | 15 | 12 | 10 | 8 | 5 |
| Differences | 6 | 5 | 5 | 4 | 3 | 2 | 1 |

When this is accomplished the apparatus is lifted out of the cylinder D and the condenser detached. A light is then applied to the fuse, and the condenser quickly replaced and plunged into the water; when gas appears through the holes at the bottom of A the time is noted, and an observation made as to the regularity of the combustion. When the combustion is at an end, which should in most cases not occupy less than 60 seconds, the stopcock of A is opened and a wire forced down the pipe to clear it; the whole is then moved up and down in the liquid with the thermometer until the temperature has attained its maximum, and the increase is then noted.

Often there is a small quantity of sample that has escaped combustion, and to make allowance for this the liquid is made acid with hydrochloric acid, and evaporated to a small bulk in a porcelain basin. The residue is filtered off, washed, and dried, and brushed into a tared crucible, dried and weighed; it is then ignited in a muffle, cooled, and again weighed; the loss is assumed to be the carbon and hydrogen unburnt.

To ascertain the temperature that this would raise the water to if completely burnt in the calorimeter: Let $V$ = volatile matter minus water in 2 grams sample, C amount of carbon and hydrogen found to be unburnt, and T, the rise in temperature

in calorimeter, then the temperature corresponding
to C will not be far short of,

$$T_2 = \frac{T_1 \times C}{V}$$

The following is an example of the determination of the heating power of a sample of Welsh
steam coal by the above process:—

Temperature of room                                    17·25° C.
   ,,         ,, water before combustion 13·20° C.
   ,,         ,,   ,,   after      ,,        20·30° C.
Increase in temperature 20·3 − 13·2   =   7·10° C.
Allowance for temperature of C and H
                                            unburnt   ·21° C.
                                                        7·31° C.
Absorption of heat by Calorimeter $\frac{1}{10}$ =   ·73° C.
                                       Total   8·04° C.

8·04 × 1,000 = 8,040 calories — that is, pounds of
water heated 1° C. by 1 pound of sample.

The latent heat of steam being 537 thermal units,
the evaporative power (lbs. of water evaporated at
100° C. (212° F.) by 1 lb. of coal) becomes $\frac{8040}{537}$ =
14·97.

On determining the heating power of substances
rich in hydrogen—such as patent and liquid fuels
—the oxygen mixture is diluted with from ·5 to 3
grams of dry kaolin clay.  The combustion is some-

times difficult to start; in most cases, however, this may be overcome by the employment of a little gunpowder or coal mixture with the fuse, the calorific value of which has been previously ascertained, and the temperature corresponding to the amount taken must be subtracted from the increase of temperature observed.

## Calculation of the Theoretical Calorific Values of Solid and Liquid Fuels from the Chemical Analysis.

Carbon and hydrogen are the only elements in solid and liquid fuels that may be considered to be the source of their heating efficiency, consequently the amount of heat that would be expected from them would be thought to depend upon the respective amounts of these elements present in the fuel ; the quantity of oxygen present, however, has to be taken into account, which is assumed to be combined with its equivalent of carbon or hydrogen as the case may be, and renders so much of the latter incapable of generating heat.

The amount of heat rendered ineffective by the quantity of oxygen present will depend upon whether the latter is supposed to be combined with carbon or hydrogen. The amount of carbon rendered latent by a given quantity of oxygen would be three times as much as the amount of hydrogen rendered latent by the same quantity of oxygen.

The heat given out in the combustion of hydrogen is always the same, but in the case of carbon it depends upon whether it is oxidised to its maximum as carbonic acid $(CO_2)$ or to its minimum carbonic oxide $(CO)$, so that carbon can have two calorific values according whether the product of combustion be CO or $CO_2$.

In practical working great loss of heat would be entailed if the carbon was not oxidised to its full, as the following shows :—

Calories.

The heat generated by the combustion of
    carbon to carbonic acid is  .    . =  8080
While the heat generated by the com-
    bustion of twice the weight of carbon
    to carbonic oxide is .   .   .   . =  4946
Loss of heat by production of CO. .   . =  3134

From accurate determinations made by Favre and Silbermann it has been found that the amount of heat generated by the combustion of hydrogen is 4·265 times as great as the heat given out by the combustion of the same weight of carbon to carbonic acid.

The relative calorific power of fuels may be deduced from the following formulæ—

(1.) Fuel containing carbon only . . . . $p = C$.

(2.) Fuel containing carbon and hydrogen . . $p = C + 4·265\,H$.

(3.) Fuel containing carbon, hydrogen, and oxygen $p = C + 4·265(H - \tfrac{1}{8}O)$.

where, p = relative calorific power

C, H and O = amounts of carbon, hydrogen, and oxygen present in 1 part of fuel.

If it be required to express the calorific power of a fuel in heat units then—

(1.) $p = 8080\ C$.

(2.) $p = 8080\ C + 34462\ H$.

(3.) $p = 8080\ C + 34462\ (H - \tfrac{1}{8}\ O)$.

The following table gives the calorific values of several substances as calculated from the above formulæ :—

| Fuel. | Composition of fuel. | | | | Relative calorific power. | Heat Units, Centigrade. | Weight of Water heated from 0° to 100° C. | Wght of Water at 100° C. converted into Steam. |
|---|---|---|---|---|---|---|---|---|
| | Carbon. | Hydrogen. | Oxygen. | Ash. | | | | |
| Hydrogen . . | — | 1·00 | — | — | 4·265 | 34,462 | 344·62 | 62·658 |
| March gas . . | 0·75 | 0·25 | — | — | 1·816 | 14,675 | 146·75 | 26·682 |
| Olefiant gas . | 0·875 | 0·143 | — | — | 1·466 | 11,849 | 118·49 | 21·543 |
| Welsh coal . . | 0·838 | 0·048 | 0·041 | 0·049 | 1·020 | 8,241 | 82·41 | 14·983 |
| Newcastle coal . | 0·821 | 0·053 | 0·057 | 0·038 | 1·017 | 8,220 | 82·20 | 14·945 |
| Carbon . . | 1·000 | — | — | — | 1·000 | 8,080 | 80·80 | 14·691 |
| Scotch coal . | 0·785 | 0·056 | 0·097 | 0·040 | 0·973 | 7,861 | 78·61 | 14·292 |
| Derbyshire coal . | 0·797 | 0·049 | 0·101 | 0·026 | 0·956 | 7,733 | 77·33 | 14·060 |
| Lancashire coal . | 0·779 | 0·053 | 0·095 | 0·049 | 0·955 | 7,717 | 77·17 | 14·031 |
| Kiln-dried peat . | 0·600 | 0·060 | 0·307 | 0·020 | 0·694 | 5,640 | 56·40 | 10·254 |
| Air-dried peat . | 0·461 | 0·046 | 0·246 | 0·015 | 0·526 | 4,250 | 42·50 | 7·727 |

In determining the thermal effect of fuels from their percentage composition, when made to burn in air, corrections have to be made for the latent heat of water produced by the combustion of the available hydrogen ; and the specific heats of the carbonic acid, water vapour, nitrogen and air have also to be taken into account. The following is a formula for arriving at the thermal effect of a fuel when completely oxidised in air :—

$$T = \frac{c\,C + c'\,H - l\,W}{S\, 3 \cdot 07\,C + 9\,H + S'\,W + S''\,N + S'''\,A}.$$

Here $T =$ increase of temperature produced by combustion.

$C$ and $H =$ quantities of carbon and hydrogen available in 1 part fuel.

$W =$ water produced by 1 part fuel.

$l =$ latent heat of water.

$S\ S'\ S''\ S''' =$ specific heats of carbonic acid, water vapour, nitrogen and air.

$C$ and $C' =$ calorific power of carbon and hydrogen.

$N =$ nitrogen in quantity of air necessary for complete combustion of fuel.

$A =$ any additional amount of air supplied for combustion.

The result obtained by the above formula expresses the highest heat attainable as compared with carbon burnt to its highest oxide under the best conditions. The amount of heat generated practically, however, is usually less than what should be obtained by calculation as above; this is due to a variety of causes, such as imperfect combustion, loss of fuel as smoke, imperfectly oxidised cinders, &c.

Rankine adopts as his unit, the weight of fuel required to evaporate 1 lb. of water at 212° F. (= 100° C.) under a pressure of 14·7 lbs. per square inch, this being equivalent to 966 British Thermal Units. The results were obtained as follows :—

Let E be the corrected and reduced evaporation,

   e the weight of water evaporated,

   $T_1$ the standard boiling point 212° F. (= 100° C.).

   $T_f$ the temperature of feed water,

   $T_b$ the actual boiling point observed;

then—

$$E = e \left\{ 1 + \frac{T_1 - T_f + 0\cdot3(T_b - T_1)}{966 \text{ F or } 537 \text{ C}} \right\}$$

The result is the number of times its own weight of water which a fuel would evaporate at the standard temperature if no loss of heat occurs; but as there always is some loss of heat, the efficiency of the furnace is expressed by the ratio, $\frac{E' \text{ (Available)}}{E \text{ Total}}$, which, if no waste occurred, would be = 1.

The loss of units of evaporation by the waste gases may be obtained by the formula,

$$\text{Loss by chimney } \frac{1 + A'}{4000} \text{ Tc (F.°)}.$$

Here $1 + A' =$ weight of burnt gas per unit weight of fuel; and Tc (F.°) the temperature of the chimney gases above that of the atmosphere. For ordinary coal $1 + A'$ ranges from 13 to 25, and for liquid fuels 16·3, if no excess of air is required.

Taking some examples with coal, with a chimney draught, the temperature of the waste gases being 600° F. $(= 315°$ C.), were,

| | | | | | | | |
|---|---|---|---|---|---|---|---|
| $1 + A'$ | . | . | . | . $=$ | 13 | 19 | 25 |
| Tc | . | . | . | . $=$ | 600° | 600° | 600° |
| Vol. of gases in cu. ft. | | . $=$ | | | 325 | 475 | 625 |
| Loss of evaporative power $=$ | | | | | 1·95 | 2·85 | 3.75 |

In estimating the evaporative efficiency of fuels from their chemical constitution, Rankine proposed the formula,

$$E = 15 \, C + 64 \, H - 8 \, O;$$

and to calculate the amount of air required for combustion, $A = 12 \, C + 36 \, H - 4\frac{1}{2} \, O$. The practical value of a fuel, however, is a little lower than estimated by the formula. The following results were obtained by Rankine :—

| | Chemical Compo-sition. | | | | | Evaporation due to | |
|---|---|---|---|---|---|---|---|
| | C. | H. | O. | A. | E. | C. | H—$\frac{o}{8}$. |
| Charcoal . . | ·93 | 0 | 0 | 11·5 | 14·0 | 14·0 | 0 |
| Coke . . . | ·88 | 0 | 0 | 10·6 | 13·2 | 13·2 | 0 |
| Rock oils } $C_{14} H_{20}$ | ·84 | ·16 | 0 | 15·75 | 22·7 | 12·7 | 10·0 |
| } $C_{26} H_{24}$ | ·85 | ·15 | 0 | 15·65 | 22·5 | 12·66 | 9·84 |
| Coal . . . | ·87 | ·05 | ·04 | 12·1 | 15·9 | 13·05 | 2·85 |
| ,, . . . | ·85 | ·05 | ·06 | 11·7 | 15·5 | 12·75 | 2·75 |
| ,, . . . | ·75 | ·05 | ·05 | 10·6 | 14·1 | 11·25 | 2·85 |
| Ethylene . . | ·75 | ·25 | 0 | 18·0 | 27·3 | 11·25 | 16·05 |
| Acetylene . . | ·85 | ·14 | 0 | 15·43 | 22·1 | 12·9 | 9·2 |
| Peat, dry . . | ·56 | ·06 | ·31 | 7·7 | 10·0 | 8·5 | 1·5 |
| Wood, dry . . | ·58 | ·05 | ·40 | 6·0 | 7·5 | 7·5 | 0 |

Dr. Paul determines the evaporative power of hydrocarbons as the sum of that of the hydrogen and carbon present, assuming that when oxidised with the theoretical proportion of air each lb. of carbon evaporates 11·359 lbs. of water at 15·5° C., and each lb. of hydrogen 41·895 lbs. of water at 15·5° F. into steam at 100° C. The results obtained by this method are given in the following table. Column 5 gives the evaporative duty when the furnace gases are discharged at 315° C. above the temperature of the air supplied to the furnace.

| | Car-bon. | Hydro-gen. | Oxy-gen. | Evaporative Power, lbs. water at 100° C. | Evaporative Duty, lbs. water at 15·5° C. |
|---|---|---|---|---|---|
| Phenol . . . . | 76·6 | 6·40 | 17·00 | 12·2437 | 10·5025 |
| Cresol . . . . | 77·7 | 7·41 | 14·82 | 13·0096 | 11·1632 |
| Naphthalin . . . | 95·75 | 6·25 | — | 15·4350 | 13·0751 |
| Anthracine . . . | 94·38 | 5·62 | — | 15·24·7 | 13·2675 |
| Xylol . . . . | 90·56 | 9·44 | — | 16·5866 | 14·2415 |
| Cumol . . . . | 90·00 | 10·00 | — | 16·7838 | 14·4126 |
| Cymol . . . . | 89·55 | 10·45 | — | 16·9422 | 14·5500 |

It is computed, generally speaking, that in average practical working 1 lb. of liquid fuel would not be likely to evaporate more than 16 lbs. of water as steam.

An example of Paul's method of obtaining the *effective* heat is as follows :—

COMBUSTION OF 1 LB. OF CARBON.

|  | Heat Units. | Equivalent Evaporation of Water. | |
|---|---|---|---|
|  |  | At 212° F. | At 60° F. |
| Total heat of combustion . . . . | 14,500 | 15 | — |
| Available heat . . . . . . . | 14,500 | — | — |
| Waste by furnace gases at 600° F. . | 3,480 | 3·6 | — |
| Effective heat . . . . . . . . | 11,020 | 11·4 | 9·8 |

COMBUSTION OF 1 LB. OF HYDROGEN.

|  | | | |
|---|---|---|---|
| Total heat of combustion . . . . | 62,032 | 64·2 | — |
| Latent heat of water vapour . . . | 8,695 | — | — |
| Available heat . . . . . . . | 53,337 | — | — |
| Waste heat of furnace gases . . . | 11,520 | 11·9 | — |
| Effective heat . . . . . . . . | 41,817 | 43·3 | 38 |

The following are the results obtained per lb. of two kinds of liquid fuel, A and B :—

A containing 86 per cent. carbon and 14 per cent. hydrogen.

B      ,,      75      ,,      ,,      25      ,,

A.

| Carbon. | Hydrogen. | Total Heat of Combustion. | Equivalent Evaporation of Water. | |
|---|---|---|---|---|
| | | | At 212° F. | At 60° F. |
| ·86 | | × 14500 = 12470 | • | |
| | ·14 | × 62032 = 8684 | | |
| | | 21154 | 21·9 | 18·8 |

| Furnace Gases. | | | Heat units in Furnace Gases. | |
|---|---|---|---|---|
| | | lbs. | | |
| Carbonic acid . | . | 3·16 | 411 | |
| Water vapour . | . | 1·26 | 359 | |
| Nitrogen . | . | 11·45 | 1683 | |
| Surplus air | . | 14·37 | 2124 | 2·2 |
| | | 30·74 | 4577 | 4·8 |

| | | | | |
|---|---|---|---|---|
| Total heat of combustion . | . | . | 21154 | |
| Latent heat of water vapour | . | . | 1217 | 1·3 |
| Available heat . | . | . | 19937 | |
| Waste in furnace gases | . | . | 4577 | 4·8 |
| Effective heat . | . | . | 15360 | 15·8 | 13·6 |
| Theoretical evaporating power . | . | . | | 21·9 |

B.

| Carbon. | Hydrogen. | Total Heat of Combustion. | Equivalent Evaporation of Water. | |
|---|---|---|---|---|
| | | | At 212° F. | At 60° F. |
| ·75 | | × 14500 = 10775 | | |
| | ·25 | × 62032 = 15508 | | |
| | | 26283 | 27·1 | 23·1 |

| Furnace Gases. | | Heat units in Furnace Gases. | | |
|---|---|---|---|---|
| | lbs. | | | |
| Carbonic acid . . | 2·75 | 358 | | |
| Water vapour . . | 2·25 | 641 | | |
| Nitrogen . . . | 13·39 | 1968 | | |
| Surplus air . . | 17·39 | 2483 | 2·6 | |
| | 35·78 | 5450 | | |

| | | | | |
|---|---|---|---|---|
| Total heat of combustion . . . | 26283 | | | |
| Latent heat of water vapour . . | 2174 | 2·2 | | |
| Available heat . . . . . | 24109 | | | |
| Waste in furnace gases . . . | 5450 | 5·6 | | |
| Effective heat . . . . . | 18659 | 19·3 | 16·6 | |
| Theoretical evaporating power . . . | | 27·1 | | |

## Gaseous Fuels.

The great progress that has recently been made in the manufacture and application of gas suitable for fuel, metallurgical and domestic heating purposes, &c., necessitates a ready method by which its analysis and heating power can be quickly deduced with fair practical accuracy.

For the very accurate and scientific analysis of complex gas mixtures, delicate processes such as Frankland and Ward's would have to be resorted to, which would, generally speaking, be far too

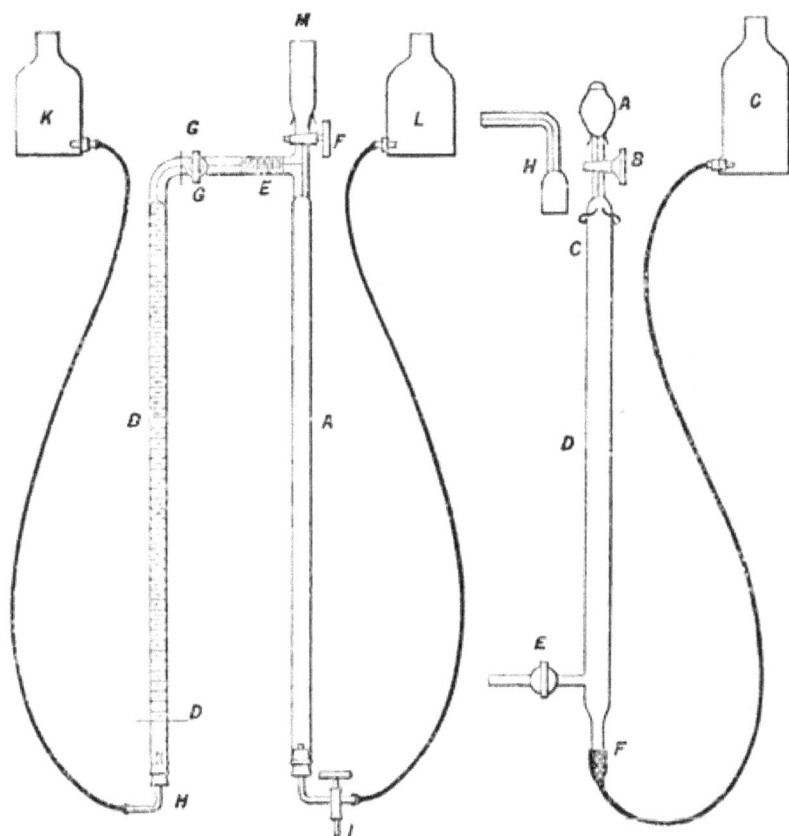

Fig. 4.

Fig. 5.

tedious and elaborate for practical working purposes. In metallurgical works, water-gas works, &c., it is often necessary to have several complete technical analyses of gas in a single day, in order

D

to give an idea of the economic working of processes, &c.

The apparatus which seems to commend itself for quick working and reasonable accuracy is that devised by Elliot, of which a sketch is here given. B, Fig. 4, is a tube graduated to 100 cc. in $\frac{1}{10}$th cc. The stop cock I is a three-way tube with a delivery tube through its stem. The bottles K and L hold about a pint each. M is a portable funnel ground to fit above F and holds 60 cc. E is a rubber tube joining A and B.

Before starting an analysis, the tubes A and B are filled with water from the bottles K and L, and manipulation of the stop-cocks C, F, and I; when water rises into the funnel M, and all air eliminated, F and G are turned off, the funnel M removed, and the tube containing the gas for analysis attached in its place; the bottle L is now slowly lowered and the stop F closed; remove the gas tube from F and replace the funnel M, raise the bottle L, and lower the bottle K, open the stopcock G, the gas is thus driven into the graduated tube B. Lower the bottle K so that the level of the water therein is in a line with the zero mark D. The gas is then adjusted to the zero mark D by the bottle L, and the stop-cock G closed, and the temperature and pressure recorded.

When the surplus gas has been displaced from

the tube A by raising L and opening F, by mani-
pulating the bottles and stop-cocks the gas is now
drawn into A; close the cocks, lower the bottle L,
and fill the funnel M with a solution of potassic
hydrate (1 in 20). Cautiously open F and allow
the liquid to flow into the tube A, always leaving,
however, about 10 cc. in M. On allowing to stand
until no further diminution in volume occurs, due
to the absorption of carbonic acid (CO) present
in the gas, the residual gas is transferred to the
tube B and measured, noting the temperature and
pressure; the loss on the original volume is $CO_2$.
Empty the tube A and wash out with water, and
refill with water as before, the gas from B is transferred to it, the funnel M is half filled with water,
and a few drops of bromine added; this is run into
the tube until fumes of bromine are seen to be
mixed with the gas.

On allowing to stand until no further contraction
occurs, due to the absorption of ethylene and other
illuminants, some of the potash used for the absorption of the $CO_2$ is added; this absorbs the
excess of bromine. When this is complete the gas
is measured as before, the loss in volume being
put down to illuminants.

After A has been cleared out and refilled with
water the gas is again brought into it as before;
the funnel M is filled with a solution of potassic

hydrate (1 in 8), to which has been added about 3 per cent. of pyrogallic acid; this is run into the tube and the gas allowed to remain over it, until any oxygen present is completely absorbed; it is then measured, loss = oxygen.   Clean out A, withdraw the gas from B into it, fill M with strong hydrochloric acid, containing 25 per cent. of cuprous chloride, and allow to stand till no further diminution in volume occurs, and measure as before; the loss is due to carbonic oxide (CO).

The residual gas may now contain marsh gas, hydrogen, and nitrogen, and to determine the proportion of these it will be necessary to explode them with oxygen by an electric spark.   This is accomplished by means of the explosion tube, Fig. 5, graduated to 100 cc. in $\frac{1}{10}$ cc. to within 2 inches from E, the stopcock B being the zero point.

The funnel A is portable as M in Fig. 4; at C there are two platinum wires fused in, which are connected with an induction coil.   The bent tube H is made to fit over the stopcock B when the funnel A is removed, which serves to facilitate the transfer of gas.   Before starting remove the funnel M in the absorption tube, and fix in its place a bent tube like H.   The gas having been transferred to A, the explosion tube is placed near it, and the bent tube H is attached to a piece of rubber tube,

long enough to reach to the corresponding bent
tube of the explosion tube.

The explosion tube is now filled with water from
the bottle G to the end of the tube H over the stopcock B. Fill the piece of tubing with water, and
connect the two bent tubes H with it. Now turn
the three-way cock I, so that the bottom of the
tube A is closed. Now open B and F, and by the
aid of the bottles draw in 20 cc. of the gas into the
explosion tube, leveling with the bottle G, and close
the stop-cocks. Connect the inlet tube E with a
gas-holder of oxygen under pressure, and intro-
duce about 20 cc. into the tube and mix; level off
and note volume. Place the bottle G below F to
expand the gases, pass the spark, and with a click
the explosion is complete. Allow the flush of heat
to debate, and observe the contraction after leveling with G.

By removing the bent tube H and fixing on the
funnel A, the amount of $CO_2$ produced can be ascertained by introducing the solution of potassic
hydrate, and observing the diminution in volume.
The formula for the calculation of the proportion of
hydrogen, marsh gas, and nitrogen present in the
20 cc. of residual gas taken (which must after-
wards be calculated on the original gas taken)
becomes :—

Let A        = original volume ;

C       = contraction ;

D       = carbonic acid formed ;

X, Y, Z = H, CH$_4$, and N respectively ;

then—

$$X = \frac{2\,C - 4\,D}{3}$$

$$Y = D$$

$$Z = \frac{3\,A - 2\,C + D}{3}$$

In working with the apparatus the analysis should be performed in a room where the temperature will remain uniform during the analysis, and care should be taken that the water, chemicals, &c., are at the same temperature as the room.

Inasmuch as a complete analysis can be made by this process in the course of an hour, the temperature and pressure, if suitable precautions be taken, would rarely be altered during the readings, and the original gas being usually saturated with moisture, no correction would be necessary for the tension of aqueous vapour, since it is measured over water.

### Calorific Value.

The heating power of a gaseous fuel containing carbonic acid, hydrogen, marsh gas, olefiant gas, nitrogen, carbonic acid, and aqueous vapour, Bunzen deduces as follows :—

$$A = 3000 \left[ x \cdot k \cdot o \cdot 57 + 1\cdot5 \cdot h \cdot 8 + 1\cdot1 \cdot g \cdot 4 + 1\cdot17 \cdot o \cdot 3\cdot43 \right] - 550 \left[ g \cdot h \cdot + 2\cdot25 \cdot g + 1\cdot29 \cdot o + w \right]$$

Where K = amount of CO

| | | |
|---|---|---|
| H = | „ | H |
| $g$ = | „ | $CH_4$ |
| $o$ = | „ | $C_2H_4$ |
| $n$ = | „ | N |
| $w$ = | „ | Aqueous vapour |
| $k$ = | „ | $CO_2$ |

For the pyrometric heating effects of gaseous fuel burnt in air the folowing formula is used :—

$$T = \frac{A}{Q(K)S + Q(w)S' + Q(n)S''}$$

$$Q(K) = K + 1\cdot57 \cdot k + 2\cdot75 \cdot g. + 3\cdot14 \cdot o$$

$$Q(w) = w + 9 \cdot h + 2\cdot25 \cdot g + 1\cdot29 \cdot o$$

$$Q(n) = n + 3\cdot33 \, (o \cdot 57 \cdot k + 8 \cdot h + 4 \cdot g + 3\cdot43 \cdot o)$$

S S'S'' = specific heats of $CO_2$, $OH_2$, and N respectively. A much higher temperature would be produced if burnt in pure oxygen, as the following calculation by Bunzen by the above formula shows :—

| | In oxygen. | In air. |
|---|---|---|
| Carbon . . . . | 9873° C. | 2458° C. |
| Carbonic oxide . . . | 7067 | 3042 |
| Olefiant gas . . . | 9187 | 5413 |
| Marsh gas . . . . | 7851 | 5329 |
| Hydrogen . . . . | 8061 | 3259 |

METHOD OF ASCERTAINING THE THERMIC VALUE OF A GAS
COMPARED WITH COAL.

*(Ford, Jour. I. and S. Inst.)*

A natural gas from the Pittsburg district has
the following average chemical composition :—

| | | | | | |
|---|---|---|---|---|---|
| Carbonic acid | . | . | . | . | 0·60 per cent. |
| Carbonic oxide | . | . | . | . | 0·60 ,, |
| Oxygen | . | . | . | . | 0·80 ,, |
| Olefiant gas | . | . | . | . | 1·00 ,, |
| Ethylic hydride | . | . | . | . | 5·00 ,, |
| Marsh gas | . | . | . | . | 67·00 ,, |
| Hydrogen | . | . | . | . | 22·00 ,, |
| Nitrogen | . | . | . | . | 3·00 ,, |

Now by the specific gravity of these gases we find
that 100 litres of this gas will weigh 64·8585 grams,
thus :—

| | | | | |
|---|---|---|---|---|
| Marsh gas | . | 67·0 litres weighs 48·0256 gms. | | |
| Olefiant gas | . | 1·0 ,, | ,, | 1·2534 ,, |
| Ethylic hydride | . | 5·0 ,, | ,, | 6·7200 ,, |
| Hydrogen | . | 22·0 ,, | ,, | 1·9712 ,, |
| Nitrogen | . | 3·0 ,, | ,, | 3·7632 ,, |
| Carbonic acid | . | ·6 ,, | ,, | 1·2257 ,, |
| Carbonic oxide | . | ·6 ,, | ,, | ·7526 ,, |
| Oxygen | . . | ·8 ,, | ,, | 1·1468 ,, |
| | Total | . | . | 64·8585 ,, |

Then if we take the heat units of these gases, we
will find

| | | | | |
|---|---|---|---|---|
| Marsh gas | . | 48·0256 grams contain 627,358 heat units. | | |
| Olefiant gas | . | 1·2534 | ,, | 14.910 ,, |
| Ethylic hydride | | 6·7200 | ,, | 77,679 ,, |
| Hydrogen | . | 1·9712 | ,, | 67,929 ,, |
| Carbonic oxide | | ·7526 | ,, | 1,808 ,, |
| Nitrogen | . | 3·7630 | ,, | — ,, |
| Carbonic acid | . | 1·2257 | ,, | — ,, |
| Oxygen | . . | 1·1468 | ,, | — ,, |
| | | 64·8585 | ,, | 789,694 ,, |

64·858 grams are almost exactly 1,000 grains, and 1 cubic foot of this gas will weigh 267·9 grains; then the 100 litres or 64·8585 grams or 1,000 grains are 3·761 cubic feet.

3·761 cubic feet of this gas contain 709,694 heat units, and 1,000 cubic feet will contain 210,069,604 heat units. Now 1,000 cubic feet of this gas will weigh 265,887 grains, or 38 lbs. avoirdupois.

We find that 64·8585 grams or 1,000 grains of carbon contain 524,046 heat units, and 265,887 grains or 38 lbs. of carbon contain 139,398,896 heat units. Then 57·25 lbs. of carbon contain the same number of heat units as 1,000 cubic feet of the natural gas, viz., 210,069,604.

Now, if we say that coke contains in round numbers 90 per cent. of carbon, then we will have 62·97 lbs. of coke equal in heat units to 1,000 cubic feet of natural gas.

Then, if a ton of coke, or 2,000 lbs., cost 10s., 62·97 lbs. will cost 4d., or 1,000 cubic feet of gas is worth 4d. for its heating power.

We will now compare the heating power of this gas with bituminous coal, taking as a basis a coal slightly above the general average of the Pittsburg coal, viz. :—

| Carbon | . | . | . | . | 82·75 per cent. |
| Hydrogen | . | . | . | . | 5·31 ,, |
| Nitrogen | . | . | . | . | 1·04 ,, |
| Oxygen | . | . | . | . | 4·64 ,, |
| Ash | . | . | . | . | 5·31 ,, |
| Sulphur | . | . | . | . | ·95 ,, |

We find that 38 lbs. of this coal contains 146,903,820 heat units. Then 54·4 lbs. of this coal contains 212,069,640 heat units or 54·4 lbs. of coal is equal in its heating power to 1,000 cubic feet of natural gas. If our coal cost us 5s. per ton of 2,000 lbs. then 54·4 lbs. will cost 1·632 pence and 1,000 cubic feet of gas will be worth for its heat units 1·632 pence.

As the price of coal increases or decreases the value of the gas will naturally vary in like proportions. Thus with the price of coal at 10s. per ton, the gas will be worth 3·264 pence per 1,000 cubic feet.

If 54·4 lbs. of coal be equal to 1,000 cubic feet of gas then 1 ton, or 2,000 lbs., will be equal to 36,764 cubic feet or 2,240 lbs. of coal will be equal to 40,768 cubic feet of natural gas.

If we compare this gas with anthracite coal, we find that 1,000 cubic feet of gas is equal to 58·4 lbs. of this coal, and 2,000 lbs. of coal is equal to 34,246 cubic feet of natural gas. Then if this coal cost 26s. per ton, 1,000 cubic feet of natural gas will be worth 9½d. for its heating power.

# TABLES

## OF

# PRACTICAL RESULTS AND ANALYSES.

COMPARATIVE CONSUMPTION OF COAL BY COMPOUND AND ORDINARY LOCOMOTIVES FOR THREE MONTHS ENDING 21ST MAY, 1886, ON THE GREAT EASTERN RAILWAY.

| Number and Class of Engines. | 1886. Four weeks ending. | Distance run. | | Coal consumed. | | |
|---|---|---|---|---|---|---|
| | | Train Miles. | Engine Miles. | Total Cwts. | lbs. per Train Mile. | lbs. per Engine Mile. |
| Eleven compounds . | 26 Mar. | 36,503½ | 37,794½ | 9,840 | 30·2 | 29·1 |
| Eleven ,, . | 23 Apl. | 34,848¼ | 35,857¾ | 8,979 | 28·8 | 28·0 |
| Eleven ,, . | 21 May | 38,117¾ | 39,102¾ | 9,939 | 29·2 | 28·4 |
| Totals and averages . | . . | 109,469½ | 112,754¾ | 28,767 | 29·4 | 28·5 |
| Six ordinary . . | 26 Mar. | 18,610 | 19,355½ | 5,724 | 34·4 | 33·0 |
| Seven ,, . | 23 Apl. | 23,761 | 24,362¼ | 7,162 | 33·7 | 32·9 |
| Seven ,, . | 21 May | 22,300 | 22,938½ | 6,661 | 33·4 | 32·5 |
| Totals and averages . | . . | 64,671 | 66,656¼ | 19,547 | 33·8 | 32·8 |
| Mean saving by compounds . . . . . | | | | | 4·4 | 4·3 |

COMPARATIVE CONSUMPTION OF COAL AND EVAPORATION OF
WATER BY COMPOUND AND ORDINARY LOCOMOTIVES WORK-
ING PASSENGER TRAINS FROM LONDON TO NORWICH ON THE
GREAT EASTERN RAILWAY.

| October, 1886. | | Compound No. 704. | Ordinary No. 565. |
|---|---|---|---|
| Coal consumption, total      . | lbs. | 2780 | 3444 |
| „         „         per mile  . | lbs. | 24·3 | 30·2 |
| Water evaporation, total   . | gallons | 2196 | 2853 |
| „         „         per lb. of coal  .      .      .      . | lbs. | 7·9 | 8·2 |
| Feed water, average per 5 minutes   .      .      .      . | gallons | 112·5 | 126·4 |
| Feed water temperature    . | Fahr. | 64° | 65° |
| Average steam pressure per square inch  .      .      . | lbs. | 138 | 122 |
| Load, London to Ipswich    . | vehicles | 14 | 15 |
| „      Ipswich to Norwich  . | vehicles | 6 | 7 |

The compound steamed freely, weather very favourable.

The ordinary engine steamed moderately, weather rather unfavour-
able.

AVERAGE COMPOSITION OF COALS FROM DIFFERENT LOCALITIES.

*(Phillips' Admiralty Coal Investigation.)*

| Locality. | Sp. Gr. | Carbon. | Hydrogen. | Nitrogen. | Sulphur. | Oxygen. | Ash. | Coke. |
|---|---|---|---|---|---|---|---|---|
| Average of 36 samples from Wales   . | 1·315 | 83·78 | 4·79 | 0·98 | 1·43 | 4·15 | 4·91 | 72·60 |
| 18 samples from New-castle   .      .      . | 1·256 | 82·12 | 5·31 | 1·35 | 1·24 | 5·69 | 3·77 | 60·67 |
| 28 samples from Lancashire  .      .      . | 1.273 | 77·90 | 5·32 | 1·30 | 1·44 | 9·53 | 4·88 | 60·22 |
| 8 samples from Scotland  . | 1·259 | 78·53 | 5·61 | 1·00 | 1·11 | 9·69 | 4·03 | 54·22 |
| 7 samples from Derby-shire   .      .      . | 1·292 | 79·68 | 4·94 | 1·41 | 1·01 | 10·28 | 2·65 | 59·32 |

## COMPOSITION OF VARIOUS ANTHRACITES.

| Locality. | Sp. Gr. | Carbon. | Hydrogen. | Oxygen, Nitrogen, and Sulphur. | Ash. | Observers. |
|---|---|---|---|---|---|---|
| Pennsylvania | 1·462 | 89·21 | 2·43 | 3·69 | 4·67 | ⎫ |
| Swansea | 1·348 | 91·29 | 2·33 | 4·80 | 1·58 | ⎪ |
| Mayenne | 1·343 | 90·20 | 4·18 | 3·37 | 2·25 | ⎬ Regnault. |
| Roidue (near Aix-la-Chapelle | 1·367 | 90·72 | 3·92 | 4·42 | 0·94 | ⎭ |
| Swansea | 1·270 | 90·58 | 3·60 | 4·10 | 1·72 | ⎫ |
| Sablé | 1·750 | 87·22 | 2·49 | 3·39 | 6·90 | ⎬ Jacquelin. |
| Vizille | 1·730 | 94·09 | 1·85 | 2·85 | 1·90 | ⎪ |
| Isere | 1·650 | 94·00 | 1·49 | 3·58 | 4·00 | ⎭ |

## TABLE SHOWING THE PROGRESSIVE DIMINUTION OF HYDROGEN AND OXYGEN FROM WOOD TO ANTHRACITE.

(*Prof. Johnson.*)

| | Carbon. | Hydrogen. | Oxygen. | Disposable Hydrogen. |
|---|---|---|---|---|
| Wood (average) | 100 | 12·18 | 83·07 | 1.80 |
| Peat ,, | 100 | 9·85 | 55·67 | 2·89 |
| Lignite (mean of 15 varieties) | 100 | 8·37 | 42·42 | 3·07 |
| Coal, South Staffordshire | 100 | 6·12 | 21·23 | 3·47 |
| Steam coal, Newcastle | 100 | 5·91 | 18·32 | 3·62 |
| Portrete in coal, S. Wales | 100 | 4·75 | 5·28 | 4·09 |
| Pennsylvanian anthracite | 100 | 2·84 | 1·74 | 2·63 |

## ANALYSIS OF PATENT FUELS.

(*Admiralty Investigation.*)

| Kind of Fuel. | Sp. Gr. | Carbon. | Hydrogen. | Nitrogen. | Sulphur. | Oxygen. | Ash. | Coke. |
|---|---|---|---|---|---|---|---|---|
| Wallch's patent fuel | 1·15 | 90·02 | 5·56 | trace | 1·62 | — | 2·91 | 85·1 |
| Livingstone's steam fuel | 1·184 | 86·07 | 4·15 | 1·80 | 1·45 | 2·03 | 4·52 | — |
| Livon's patent fuel | 1·13 | 86·36 | 4·56 | 1·06 | 1·29 | 2·07 | 4·66 | — |
| Wylam's ,, | 1·10 | 79·91 | 5·69 | 1·68 | 1·25 | 6·63 | 4·84 | 65·8 |
| Bell's ,, | 1·14 | 87·88 | 5·22 | 0·81 | 0·71 | 0·42 | 4·96 | 71·7 |
| Holland & Green's patent fuel | 1·302 | 70·14 | 4·65 | 1·15 | — | — | 13·73 | — |

AVERAGE VALUE OF COALS FROM DIFFERENT LOCALITIES.

| Locality. | Lbs. of water evaporated from 100° C. by 1 lb. of coal. | Number of lbs. evaporated per hour. | Weight in lbs. of 1 cubic foot of coal as used for fuel. | Space occupied by 1 ton in cubic feet. | Results obtained in experiments on cohesive power of coals (per centage of large coals). | Per cent. of sulphur in coals. |
|---|---|---|---|---|---|---|
| Average of 37 samples from Wales . . | 9·05 | 448·2 | 53·1 | 42·71 | 60·9 | 1·42 |
| Average of 17 samples from Newcastle . | 8·37 | 411·1 | 49·8 | 45·30 | 67·5 | 0·94 |
| Average of 28 samples from Lancashire . | 7·94 | 447·6 | 49·7 | 45·15 | 73·5 | 1·42 |
| Average of 8 samples from Scotland . | 7·70 | 431·4 | 50·0 | 49·99 | 73·4 | 1·45 |
| Average of 8 samples from Derbyshire . | 7·58 | 432·7 | 47·2 | 47·45 | 80·9 | 1·01 |

COMPARATIVE VALUES OF WELSH STEAM COALS.

(*Portsmouth Dockyard Experiments.*)

| | Lbs. of water evaporated by 1 lb. of coal. | Percentage of clinker and ash. |
|---|---|---|
| Nixon's Navigation . . | 10·05 | 5 37 |
| Wayne's Merthyr . . | 10·05 | 5·37 |
| Thomas ,, . . | 9·79 | 5·47 |
| Naubudyn ,, . . | 9·62 | 5·48 |
| Ynsfaio . . . | 9·52 | 6·76 |
| Merthyr Dare . . | 9·45 | 5·48 |
| Resolven Merthyr . . | 9·41 | 6·04 |
| Insoles . . . . | 9·37 | 6·52 |
| Averages . . | 9.65 | 5·81 |

RESULTS OF EVAPORATIVE DUTIES OF NEWCASTLE AND WELSH COALS IN THE COAL-TESTING MARINE BOILER AT KEYHAM STEAM FACTORY.

| Coal. | Area of Fire-grate. | Coal consumed per hour. | Coal per square foot of grate, per hour. | Water consumed from 100° per hour. | Water per sq foot of grate per hour. | Water evaporated from 100°C. per lb. of coal. |
|---|---|---|---|---|---|---|
| | sq. ft. | cwts. | lbs. | cub. ft. | cub. ft. | lbs. |
| FIRST SERIES. | | | | | | |
| (With common doors.) | | | | | | |
| Welsh, Wayne's Merthyr, Resolven, Merthyr Dare, Gellia, Ca loston . . | 14·0 | 1·93 | 15·44 | 32·4 | 2·31 | 10·42 |
| Hartley Main Newcastle . | 14·0 | 2·32 | 18·56 | 34·5 | 2·46 | 9·22 |
| ½ Welsh ½ Hartley . . | 14·0 | 1·92 | 15·40 | 30·4 | 2·17 | 9·81 |
| 2 ,, 1 ,, . . | 14·0 | 1·76 | 14·08 | 28·7 | 2·05 | 10·12 |
| 1 ,, 2 ,, . . | 14·0 | 1·96 | 15·70 | 30·7 | 2·20 | 9·72 |
| (With perforated doors.) | | | | | | |
| Hartley Main . . | 14·0 | 2·06 | 16·50 | 30·2 | 2·16 | 9·10 |
| SECOND SERIES. | | | | | | |
| (With common doors.) | | | | | | |
| Welsh—Powell's Duffryn, Nixon's Navigation, Davis's Merthyr . | 14·0 | 2·09 | 16·68 | 37·1 | 2·65 | 11·05 |
| Newcastle — Davidson's Hartley, Hasting's Hartley . | 14.0 | 2·29 | 18·29 | 34·5 | 2·46 | 9·39 |
| ½ Welsh ½ Hartleys . . | 14·0 | 2·03 | 16·24 | 34·4 | 2·46 | 10·56 |
| 2 ,, 1 ,, . | 14·0 | 2·43 | 16·36 | 35·0 | 2·50 | 10·61 |
| Welsh . . . | 14·0 | 2·19 | 17·48 | 39·3 | 2·80 | 11·16 |
| THIRD SERIES. | | | | | | |
| (With perforated doors.) | | | | | | |
| Welsh coal . . . | 14·0 | 1·87 | 14·95 | 32·7 | 2·34 | 10·86 |
| Hartleys . . . | 14·0 | 2·13 | 17·04 | 32·8 | 2·34 | 9·61 |
| ½ Welsh ½ Hartleys . . | 14·0 | 2·18 | 17·44 | 37·1 | 2·65 | 10·54 |
| 2 ,, 1 ,, . | 14·0 | 2·08 | 16·64 | 35·7 | 2·55 | 10·64 |
| 1 ,, 2 ,, . | 14·0 | 2·18 | 17·42 | 36·5 | 2·61 | 10·39 |
| David on's Hartley . . | 14·0 | 2·86 | 22·88 | 42·9 | 3·06 | 9·31 |
| ½ Hartley ½ Welsh . . | 14·0 | 2·30 | 18·40 | 31·0 | 2·22 | 10·80 |
| FOURTH SERIES. | | | | | | |
| (With smaller grate area. With common doors.) | | | | | | |
| Welsh coal . | 10·5 | 2·11 | 22·46 | 38·3 | 3·65 | 11·31 |
| ½ Welsh small ½ Davidson's Hartley . | 10·5 | 2·02 | 21·60 | 36·0 | 3·43 | 11·06 |
| ½ Welsh beans ½ Hasting's Hartley . . . | 10·5 | 2·14 | 22·85 | 36·7 | 3·50 | 10·65 |
| FIFTH SERIES. | | | | | | |
| (With perforated doors.) | | | | | | |
| Hartleys . . . | 10·5 | 2·29 | 24·40 | 42·0 | 4·00 | 11·42 |
| ½ Welsh ½ Davidson's Hartley . . . | 10·5 | 2·10 | 22·34 | 39·3 | 3·74 | 11·65 |

DISPOSAL OF HEAT OF COMBUSTION OF COAL
BURNT UNDER STEAM BOILERS, AND COMPOSI-
TION OF THE PRODUCTS OF COMBUSTION.

The following are results obtained by MM.
Scheurer-Kestner and Meunier in 1868 in a French
boiler at Thann. (*Vide Bulletin de la Société In-
dustrielle de Mulhouse.*)

|  | sq. ft. |
|---|---|
| Heating surface of heaters    .    .    . | 301·3 |
| „           „          boiler .    .    .    . | 129·1 |
| Total of boiler    . | 430·4 |
| Heating surface of feed heaters    .    . | 764·0 |
|  | 1,194·4 |

| | |
|---|---|
| Direct heating surface exposed to fire    . | 32·3 |
| Area of fire-grate    .    .    .    .    . | 19·3 |
| Area of air-spaces through grate    .    . | 5.5 |
| Ratio of grate area to total surface of boiler    .    .    .    .    .    .    . | 1 to 22·3 |
| Ratio of grate area to total surface of boiler and fuel heaters    .    .    .    . | 1 to 63 |

The average composition of the Ronchamp coal
used in the experiments devoted to the analysis of
the chimney gauges was as follows :—

| Carbon | . | . | . | . | . | . | 70 |
|--------|---|---|---|---|---|---|------|
| Hydrogen | . | . | . | . | . | . | 4 |
| Oxygen | . | . | . | . | . | . | 4 |
| Nitrogen | . | . | . | . | . | . | 1 |
| Ash | . | . | . | . | . | . | 21 |
| | | | | | | | 100 |

ANALYSIS OF THE PRODUCT OF COMBUSTION OF RONCHAMP
COAL UNDER A FRENCH BOILER AT THANN.

| No. of Experiments. | Coal per sq. ft. grate per hour. | Weight of each Charge. | Intervals of Charges. | Composition of the Gases. | | | | | | Total Air per lb. of Coal. |
|---|---|---|---|---|---|---|---|---|---|---|
| | | | | Carbonic Acid. | Carbonic Oxide. | Carbon Vapour. | Hydrogen. | Nitrogen. | Free Air. | |
| | lbs. | lbs. | mins. | p. c. | p. c. | p. c. | p. c. | p. c. | p. c. | cu. ft. |
| 12 | 8·2 | 15 4 | 5 | 14·9 | ·84 | 1·15 | 1·35 | 75·1 | 6·7 | 110·0 |
| 11 | 9·6 | 30·8 | 8 | 14·2 | ·97 | 1·11 | 1·11 | 72·5 | 10·5 | 116·2 |
| 9 | 9·6 | 15·4 | 4 | 14·6 | ·86 | ·56 | ·56 | 70·1 | 13·3 | 134·7 |
| 14 | 8·2 | 30·8 | 10 | 13·4 | ·24 | 1·41 | 1·41 | 63·7 | 20·9 | 144·5 |
| 13 | 8·2 | 15·4 | 5 | 13·3 | — | ·91 | ·91 | 67·7 | 17·6 | 147·5 |
| 8 | 4·7 | 15·4 | 8 | 12·9 | — | ·96 | ·96 | 59·7 | 26·2 | 156·6 |
| 10 | 19·0 | 15·4 | 2 | 10·9 | — | ·19 | ·19 | 45·9 | 42·8 | 198·3 |
| 7 | 3·4 | 13·2 | 10 | 8·2 | — | 0·4 | ·52 | 37·4 | 53·8 | 260·5 |

CARBON AND HYDROGEN IN ESCAPED COMBUSTIBLE GASES.

| No. of Experiments. | Per cent. of Total Carbon. | | | Per cent. of Total Hydrogen. | Temperature of Gases leaving Feed heaters. |
|---|---|---|---|---|---|
| | In Oxide. | In Hydro-carbons. | Total. | | |
| 12 | 4 1 | 11·4 | 15·6 | 19·5 | 119° C. |
| 11 | 5·0 | 10·2 | 15·3 | 16·7 | 128 |
| 9 | 5·2 | 5·9 | 11·2 | 9·9 | 126 |
| 14 | 1·5 | 4·1 | 5·7 | 26·9 | — |
| 13 | — | — | 6·1 | 17·5 | 135 |
| 8 | — | — | 3·9 | 19·7 | 93 |
| 10 | — | — | 3·4 | 4·7 | 156 |
| 7 | — | — | ·9 | 17·7 | 94 |

E

It appears from these results that the most effective combination is arrived at when about one-third of the gaseous products consists of free air.

## Influence of Excess of Air.

The following table shows the influence of excess of air in using Friedrichsthal and Altenwald coal :—

| Friedrichsthal Coal. | | Altenwald Coal. | |
|---|---|---|---|
| Free Air. Per cent. | Lbs. Water Evap. per lb. of Coal. | Free Air. Per cent. | Lbs. Water Evap. per lb. of Coal. |
| 40 | 6·80 | 35 | 7·06 |
| 36 | 6·46 | 33 | 7·28 |
| 30 | 6·38 | 32 | 7·02 |
| 27 | 6·19 | 30 | 6·79 |
| 27 | 6·23 | 28 | 6·85 |
| 24 | 5·68 | 25 | 6·71 |
| 23 | 5·80 | 23 | 6·66 |

## Distribution of the Heat of Combustion.

For the comparison of the absolute heat of combustion of coals, the boiler, firegrate, &c., in this case had the following dimensions :—

sq. ft.

Area of fire-grate 4·6 feet long by 4·47 feet wide .    .    .    .    .    .    .    . 20·6

Ratio of grate to heating surface of boiler . 1 to 21

„    to total surface of boiler and feed heaters    .    .    .    .    .    . 1 to 5

|  | cu. ft. |
|---|---|
| Total capacity of boiler   .    .    .    . | 423·6 |
| „    „    feed heaters   .   .   . | 317·7 |
| Water room in boiler  .    .    .    .    . | 335·3 |
| Steam room in boiler  .    .    .    .    . | 88·3 |

|  | sq. ft. |
|---|---|
| Heated surface of brickwork for conduction and radiation of heat   .   .   above | 1,290 |

The following are the results obtained with various coals in the above boiler :—

| Coal. | Observed Total Heat of Combustion. Units. | Air at 17° C. per lb. of Coal. Cu. ft. | Free Air. Per cent. | Temperatures. | | | Ash. | Pressure of Steams. Atmosphere. | Water per lb. of Coal from and at 100° C. |
|---|---|---|---|---|---|---|---|---|---|
| | | | | Air. °C. | Feed Heater. °C. | Smoke. °C. | | | |
| | | | | | | | | | lbs. |
| Ronchamp, No. 3   . | 7,825 | 152·3 | 24·7 | 17 | 65 | 132 | 17·3 | 4·46 | 8·77 |
| „    No. 4   . | 7,775 | 160·7 | 29·1 | 21 | 71 | 138 | 15·8 | 4·84 | 9·49 |
| Sarrebrük (mean of 7 coals)  .   .   . | 7,500 | 159·0 | 31·0 | 20 | 72 | 130 | 14·0 | 4·72 | 8·17 |
| Blanzy, Montceau  . | 7,067 | 135·4 | 23·0 | 18 | 65 | 160 | 12·0 | 4·60 | 7·89 |
| „    anthracite  . | 7,125 | 152·3 | 30·5 | 16 | 64 | 175 | 24·4 | 4·58 | 8·18 |
| Creusot, anthracite  . | 8,949 | 269·0 | 47·6 | — | 72 | 144 | 9·1 | 4·60 | 10·53 |
| Creusot, ⅔ ;  Ronchamp, ⅓   .   .   . | 8,565 | 230·1 | 36·2 | 11 | 67 | 132 | 13·4 | 4·71 | 10·54 |
| Creusot, ⅘ ;  Ronchamp, ⅕.   .   . | 8,630 | 214·9 | 34·2 | 8 | 62 | 145 | 15·9 | 4·71 | 9·83 |
| Wood, charcoal  . | 8,080 | 250·. | 42·5 | — | — | 155 | 0·5 | — | 9·2 |

The absolute heat of combustion of the fuels was estimated from the above to have been distributed about as follows:

|                                                               | per. cent. |
|---------------------------------------------------------------|-----------|
| Heat in the steam (about 60 lbs. pressure) .                  | 61·0      |
| Heat ungenerated in the combustible gases                     | 5·5       |
| Heat lost in the clinker and ash .    .    .                   | 1·5       |
| Heat carried off in the gaseous products of combustion   .    .    .    .    .    . | 5·5 |
| Heat ungenerated in the smoke carbon    .                     | ·5        |
| Heat absorbed in the evaporation of the hygrometric water, and water newly formed  .    .    .    .    .    .    . | 2·5 |
| Heat lost in the brickwork .    .    .    .                    | 23·5      |
| Total    .    .    .    .                                      | 100·0     |

## Results obtained with Liquid Fuel by Mr. James Holden, M.I.C.E., on the Great Eastern Railway.

### Experiment on a small Cornish Boiler.

#### *Coal only used.*

1887. Consumption during one week from August 15th to 20th (inclusive), 74¼ hours' work, including lighting up = 80½ cwt. = 121·3 lbs. per hour.

Cost for 100 hours = 12,130 lbs. of coal at 11s. per ton = £2 19s. 7½d.

#### *Coal, Coke and Tar used. "Holden's System."*

1888. Consumption during one week from June

25th to 30th (inclusive), 87¾ hours' working; including lighting up

$$= \text{coal } 15 \text{ cwt.} = 19\cdot2 \text{ lbs. per hour}$$
$$= \text{coke } 11\tfrac{1}{2}\text{cwt.} = 14\cdot7 \quad ,, \qquad ,,$$
$$\text{Gas tar } 280 \text{ gallons} = 35\cdot1 \quad ,, \qquad ,,$$
$$\text{Total } 69\cdot0 \quad ,, \qquad ,,$$

Cost for 100 hours

|  |  | £ | s. | d. |
|---|---|---|---|---|
| = 1,920 lbs. of coal at 11s. per ton | = | 0 | 9 | 5½ |
| = 1,470 lbs. of coke at 9s. 6d.   ,, | = | 0 | 6 | 1½ |
| = 3,510 lbs. of tar at 12s. 6d. ,, | = | 0 | 19 | 7¼ |
| Total | | £1 | 15 | 2 |

COMPARATIVE COST OF LIQUID FUEL AND COAL ON LOCOMOTIVES OF THE GREAT EASTERN RAILWAY.

| No. of Engine. | Total Miles run. | Total lbs. used. | | | Lbs. per Mile. | | | Total lbs. of Coal, Liquid Fuel, and Chalk, per mile. | Proportion of Liquid Fuel and Chalk to Coal used. | Total cost of Fuel. | | | Cost per mile in Pence. |
|---|---|---|---|---|---|---|---|---|---|---|---|---|---|
| | | Coal. | Liquid Fuel. | Chalk. | Coal. | Fuel. | Chalk. | | | £ | s. | d. | |
| | | | | | | | | | per cent. | | | | |
| 193 | 951¾ | 13,511 | 10,505 | 784 | 14·2 | 11·c | 8 | 26·0 | 83 | 9 | 1 | 5 | 2·28 |
| 194 | 951¼ | 27,738 | — | — | 29·1 | — | — | 29·1 | — | 9 | 4 | 10½ | 2·33 |
| Difference in favour of No. 193 engine . . | | | | | | | | | | 0 | 3 | 5½ | ·05 |

N.B.   Cost of coal computed at 14/11 per ton (Radford coal used).
   ,,   liquid fuel computed at 1⅛d. per gallon of 11 lbs.
   ,,   chalk computed at 5/6 per ton.

COMPARATIVE EVAPORATION OF WATER BY DIFFERENT LIQUID FUELS INJECTED AND SPRAYED WITH HOLDEN'S PATENT INJECTOR.

| ON A SMALL VERTICAL BOILER ABOUT 6-H.P. WORKING AT 50 LBS. PER SQUARE INCH PRESSURE. | Evaporative power of Liquid Fuel alone. |
|---|---|
| With Yorks coal                          4·7 lbs. water per lb. of fuel | |
| ,,    ,,    and coal tar   5·7   ,,    ,,    ,, | 6 lbs. |
| ,,    ,,    and green oil 6·6   ,,    ,,    ,, | 7·3 ,, |
| ,,    ,,    and astatki   9·1   ,,    ,,    ,, | 10·2 ,, |
| **ON A SMALL CORNISH BOILER ABOUT 30-H.P. WORKING AT A PRESSURE OF 35 LBS. PER SQUARE INCH.** | |
| With Yorks coal                          8·4 lbs. water per lb. of fuel | |
| ,,    ,,    and coal tar   9·9   ,,    ,,    ,, | 11·3 lbs. |
| ,,    ,,    and green oil 10·3   ,,    ,,    ,, | 12·3 ,, |
| ,,    ,,    and astatki   12·3   ,,    ,,    ,, | 14·5 ,, |
| **ON A LOCO. BOILER WORKING AS STATIONARY AT 80 LBS. PER SQUARE INCH PRESSURE.** | |
| With Yorks coal                          9·1 lbs. water per lb. of fuel | 15·1 lbs. |
| ,,    ,,    and green oil 12·9   ,,    ,,    ,, | |

In all cases the figures give evaporation from feed water at atmospheric temperature.

Shale oil about equals green oil for evaporative value.

COMPARATIVE CONSUMPTION OF COAL AND OIL IN ORDINARY AND COMPOUND PASSENGER LOCOMOTIVES ON THE BUENOS AYRES AND ROSARIO RAILWAY.

| Four weeks from 27 May to 25 June, 1887. | Distance run. | Coal, lbs. | | Oil, lbs. | |
|---|---|---|---|---|---|
| | | Total. | Per Mile. | Total. | Per 100 Miles. |
| Ordinary engine, No. 10. | 3,937 | 111,001 | 28·19 | 595·24 | 15·12 |
| Compound ,,    No. 34. | 3,638 | 81,350 | 22·36 | 599·65 | 16·48 |

Thus the compound burnt 20⅝ per cent. less coal than the ordinary engine, but used 9 per cent. more oil; the actual money saving was 3 dollars 20 cents per hundred miles. The working pressure in the ordinary engine was 150 lbs. per square inch, and in the compound 160 lbs.

The following are very instructive results obtained by Mr. Urquhart on the Grazi and Tsaritsin Railway locomotives, showing the efficiency of petroleum refuse.

COMPARATIVE TRIALS WITH PETROLEUM, ANTHRACITE, BITUMINOUS COAL, AND WOOD, BETWEEN ARCHEDA AND TSARITSIN ON GRAZI AND TSARITSIN RAILWAY IN WINTER.

| Train alone. | | Dis-tance run. | Car. Miles. | Fuel. | Consumption, including Lighting-up. | | Cost of Fuel per Train Mile.* | Atmospheric Temperature and Weather. |
| No. of Loaded Cars. | Gross Load. | | | | Total. | Per Train Mile. | | |
|---|---|---|---|---|---|---|---|---|
| | tons. | miles. | | | lbs. | lbs. | pence. | |
| 25 | 400 | 388 | 9,700 | Anthracite | 31,779 | 81·90 | 11·957 | —14° to |
| 25 | 400 | 388 | 9,700 | { Bituminous coal . } | 37,557·5 | 96·53 | 14·093 | —13° C. Strong side |
| 25 | 400 | 194 | 4,850 | { Petroleum refuse } | 9,462 | 48·77 | 5·487 | wind. |
| 25 | 400 | 194 | 4,850 | Anthracite | 12,639·5 | 65·15 | 9·512 | —6° to |
| 25 | 400 | 194 | 4,850 | { Wood in billets } | Cubic ft. 1,071·8 | Cu. ft. 5·52 | 8·5 | —11° C. Light side |
| 25 | 400 | 194 | 4,850 | { Petroleum refuse } | Lbs. 7,223 | Lbs. 37·23 | 4·188 | wind. |

* For prices of fuel, see next page.

The following are the additional data supplied :—

Prices of Fuel :—Petroleum refuse, 21s. per ton.

Anthracite and bituminous coal, 27s. 3d. per ton.

Wood, 1·47d. per cubic foot.

Dimensions of Locomotives :—Cylinders, 18⅛″ dia. and 24″ stroke. Wheels, 4 ft. 3 in. dia. Total heating surface, 1,248 sq. ft. Total adhesion weight, 36 tons. Boiler pressure, 8 to 9 atmos.

The following results were obtained on the above line in summer time :—

| Train alone. | | Train Miles. | Fuel. | Consumption, including Lighting-up. | | Cost of Fuel per Train Mile. |
|---|---|---|---|---|---|---|
| No. of Loaded Cars. | Gross Load. | | | Total. | Per Train Mile. | |
| No. | Tons. | | | lbs. | lbs. | pence. |
| 30 | 480 | 194 | Bituminous coal | 14,084·07 | 72·598 | 10·599 |
| 30 | 480 | 194 | Petroleum refuse | 6,175·325 | 31·831 | 3·581 |
| 30 | 480 | 194 | Anthracite    . | 12,784·002 | 65·897 | 9·621 |
| 30 | 480 | 194 | Petroleum refuse | 6,103·097 | 31·459 | 3·539 |

## COMPARATIVE MONTHLY AVERAGES DURING 1883 WITH COAL AND PETROLEUM REFUSE IN LOCOMOTIVES, WORKING MAIN LINE TRAINS ON GRAZI AND TSARITSIN RAILWAY.

### CONSUMPTION OF FUEL PER TRAIN MILE.

| Locomotives. | Trains. | Fuel. | Monthly Averages of Consumption per Train Mile in Lbs. | | | | | | | | | | | | Mean. |
|---|---|---|---|---|---|---|---|---|---|---|---|---|---|---|---|
| | | | Jan. | Feb. | Mar. | April. | May. | June. | July. | Aug. | Sept. | Oct. | Nov. | Dec. | |
| Eight wheels coupled | Goods | Coal | 98·06 | 108·96 | 100·79 | 76·27 | 76·27 | 79·00 | 74·91 | 73·55 | 79·00 | 85·81 | 78·06 | 95·34 | 87·17 |
| Six wheels coupled | Goods | Coal and Petroleum refuse | 73·55 | 77·63 | 70·82 | 64·01 | 55·84 | 61·29 | 54·48 | 55·84 | 65·38 | 80·36 | 92·62 | 85·81 | 69·80 |
| Four wheels coupled | Mixed | Coal | 53·12 | 54·48 | 46·31 | 42·22 | 34·05 | 35·41 | 31·33 | 36·10 | 40·86 | 39·50 | 50·39 | 54·48 | 43·19 |
| Four wheels coupled | Passenger | Coal | 51·76 | 76·27 | 43·58 | 34·05 | 36·77 | 35·41 | 42·22 | 49·03 | 51·76 | 40·86 | 49·93 | 58·57 | 47·44 |
| | | Coal and Petroleum refuse | 40·86 | 49·03 | 46·31 | 36·77 | 34·05 | 32·69 | 31·33 | 32·69 | 36·77 | 39·50 | 42·22 | 50·39 | 39·38 |
| | | Coal and Petroleum refuse | — | — | — | — | — | — | — | — | 20·43 | 31·33 | 32·69 | 34·05 | 29·62 |

### COST OF FUEL PER TRAIN MILE.

| Locomotives. | Trains. | Fuel. | Monthly Averages of Cost per Train Mile in Pence. | | | | | | | | | | | | Mean. |
|---|---|---|---|---|---|---|---|---|---|---|---|---|---|---|---|
| | | | Jan. | Feb. | Mar. | April. | May. | June. | July. | Aug. | Sept. | Oct. | Nov. | Dec. | |
| Eight wheels coupled | Goods | Coal | 13·495 | 15·635 | 14·346 | 11·226 | 11·316 | 11·704 | 11·186 | 10·925 | 11·552 | 12·471 | 14·600 | 13·930 | 12·699 |
| Six wheels coupled | Goods | Coal and Petroleum refuse | 10·520 | 11·099 | 9·897 | 9·405 | 8·387 | 9·317 | 8·217 | 8·279 | 9·694 | 11·776 | 13·047 | 12·344 | 10·212 |
| Four wheels coupled | Mixed | Coal | 7·294 | 7·602 | 6·158 | 4·973 | 4·040 | 4·163 | 3·617 | 4·170 | 4·948 | 4·771 | 6·817 | 7·356 | 5·495 |
| Four wheels coupled | Passenger | Coal | 7·355 | 10·973 | 6·672 | 4·934 | 5·464 | 4·988 | 6·267 | 7·269 | 11·215 | 5·908 | 7·102 | 8·652 | 6·932 |
| | | Coal and Petroleum refuse | 6·070 | 7·023 | 6·602 | 5·144 | 4·825 | 4·670 | 4·445 | 4·819 | 5·165 | 5·633 | 6·111 | 7·580 | 5·672 |
| | | Coal and Petroleum refuse | — | — | — | — | — | — | — | — | 2·411 | 5·611 | 4·366 | 4·634 | 3·808 |

Of the Coal consumed 49 per cent. was anthracite and 51 per cent. was bituminous coal.

Mean Price of Fuel:—Petroleum refuse, 23s. 4d. per ton; Coal, 27s. per ton.

## Theoretical Evaporative Value of Petroleum Fuel and Coal.

| Fuel. | Specific Gravity at 32° F. | Chemical Composition. | | | | Heating Power. British Thermal Units. | Theoretical Evaporation. Lbs. of Water per lbs. of Fuel. | |
|---|---|---|---|---|---|---|---|---|
| | | Carbon, per Cent. | Hydrogen, per Cent. | Oxygen, per Cent. | Sulphur, per Cent. | | From and at 212° F. | At 8½ Atmospheric effective Pressure. |
| | | | | | | | lbs. | lbs. |
| Pennsylvanian heavy crude oil | ·886 | 84·9 | 13·7 | 1·4 | — | 20,736 | 21·48 | 17·8 |
| Caucasian light crude oil . | ·884 | 86·3 | 13·6 | 0·1 | — | 22,027 | 22·79 | 18·9 |
| Caucasian heavy crude oil . | ·938 | 86·6 | 12·3 | 1·1 | — | 20,138 | 20·85 | 17·3 |
| Petroleum refuse . | ·928 | 87·1 | 11·7 | 1·2 | — | 19,832 | 20·53 | 17·1 |
| Good English coal, mean of 98 samples . . | 1·380 | 80·0 | 5·0 | 8·0 | 1·25 | 14,112 | 14·61 | 12·16 |

In comparing petroleum refuse (Russian) and anthracite, Mr. Urquhart says that "the former has a theoretical evaporative power of 16·2 lbs. of water per lb. of fuel, and the latter of 12·2 lbs., at an effective pressure of 8 atmospheres, or 120 lbs. per square inch, hence petroleum has, weight for weight, 33 per cent. higher evaporative value than anthracite. Now in locomotive practice, a mean evaporation of from 7 to 7½ lbs. of water per lb. of anthracite, is about what is generally obtained, thus giving about 60 per cent. of efficiency, while 40 per cent. of the heating power is unavoidably lost. But with petroleum an evaporation of 12·25 lbs. is practically obtained, giving $\dfrac{12\cdot25}{16\cdot2} = 75$ per cent. efficiency. Thus, in the first place petroleum is theoretically 33 per cent. superior to anthracite in evaporative power, and, secondly, its useful effect is 15 per cent. greater, being 75 per cent. instead of 6 per cent., while thirdly, weight for weight, the practical evaporative value of petroleum must be reckoned as at least from $\dfrac{12\cdot25 - 7\cdot50}{7\cdot50} = 63$ per cent. to $\dfrac{12\cdot25 - 7\cdot00}{7\cdot00} = 75$ per cent. higher than that of anthracite.

## Analyses of Various Gaseous Fuels.

### ANALYSIS OF SIEMENS PRODUCER GAS.

(*Trans. Amer. Inst. Min. Eng.*)

|  | 1 | 2 | 3 | 4 | 5 |
|---|---|---|---|---|---|
|  | per cent. | per cent. | per cent. | per cent. | per cent. |
| Carbonic acid . . . | 3·9 | 8·6 | 9·3 | 1·5 | 6·1 |
| Carbonic oxide . . | 27·3 | 20·0 | 16·5 | 23·6 | 22·3 |
| Hydrogen . . . . | — | 8·7 | 8·6 | 6·0 | 28·7 |
| Marsh gas . . . . | 1·4 | 1·2 | 2·7 | 3·0 | 1·0 |
| Nitrogen . . . . | 67·4 | 61·4 | 62·9 | 65·9 | 41·9 |
| Heat units . . . . | 93966 | 97184 | 99074 | 114939 | 164164 |

### ANALYSIS OF AMERICAN NATURAL GAS.

*(Ford, Jr. I. and S. Inst.)*

|  | 1 | 2 | 3 | 4 | 5 |
|---|---|---|---|---|---|
|  | per cent. | per cent. | per cent. | per cent. | per cent. |
| Carbonic acid . . . | 0·8 | 0·6 | nil. | 0·4 | nil. |
| Carbonic oxide . . | 1·0 | 0·8 | 0·58 | 0·4 | 1·00 |
| Oxygen . . . . . | 1·1 | 0·8 | 0·78 | 0·8 | 2·10 |
| Olefiant gas . . . | 0·7 | 0·8 | 0·98 | 0·6 | 0·80 |
| Ethylic hydride . . | 3·6 | 5·5 | 7·92 | 12·3 | 5·20 |
| Marsh gas . . . . | 72·18 | 65·25 | 60·70 | 49·58 | 57·85 |
| Hydrogen . . . . | 20·02 | 26·16 | 29·03 | 35·92 | 9·64 |
| Nitrogen . . . . | nil. | nil. | nil. | nil. | 23·41 |
| Heat units . . . . | 728746 | 698852 | 627170 | 745813 | 592380 |

### ANALYSIS OF THE GASES AT DIFFERENT DEPTHS OF THE ALFRETON BLAST FURNACE.

|  | Distance below the Furnace Mouth. | | | | |
|---|---|---|---|---|---|
|  | 8 feet | 14 feet | 20 feet | 24 feet | 34 feet |
| Nitrogen . . . . | 54·77 | 50·95 | 60·46 | 56·75 | 58·05 |
| Carbonic acid . . . | 9·42 | 9·10 | 10·83 | 10·08 | — |
| Carbonic oxide . . | 20·24 | 19·32 | 19·48 | 25·19 | 37·43 |
| Marsh gas . . . . | 8·23 | 6·64 | 4·40 | 2·33 | — |
| Hydrogen . . . . | 6·49 | 12·42 | 4·83 | 5·65 | 3·18 |
| Olefiant gas . . . | 0·85 | 1·57 | — | — | — |
| Cyanogen . . . . | — | — | — | trace | 1·34 |
|  | 100·00 | 100·00 | 100·00 | 100·00 | 100·00 |

ANALYSIS OF WATER-GAS.

| | (1) "Strong" Gas (Dr. Moore). | (2) Water-Gas Coke used (Langlois). | (3) Water-Gas Coke used (Frankland). | (4) Gas made by Beilby's Process. |
|---|---|---|---|---|
| Carbonic acid . . . | 2·05 | 12·000 | 13·80 | 21·32 |
| Carbonic oxide . . | 35·88 | 31·860 | 29·30 | 10·72 |
| Marsh gas . . . . | 4·11 | 1·62 ⎱ | ⎰ 56·9 | nil. |
| Hydrogen . . . . | 52·76 | 54·52 ⎰ | ⎱ | 37·19 |
| Oxygen . . . . . | 0·77 | — | — | |
| Nitrogen . . . . | 4·43 | — | — | 30·77 |
| | 100·00 | 100·00 | 100·0 | 100·00 |

GASES OCCLUDED IN COAL.

(*Thomas.*)

| Sample of Coal. | Gas evolved by 100 grams at 100° C. in a vacuum cc. | $CO_2$. | $CH_4$. | $C_2H_6$. | $(C_2H_5)_2$. | N. |
|---|---|---|---|---|---|---|
| Wigan cannel, 5/3 seam 350 yards deep . . | 421·3 | 6·44 | 80·69 | 4·75 | — | 8·12 |
| Do., 3/2 seam, 600 yards deep. . . . | 350·6 | 9·05 | 77·19 | 7·80 | — | 5·96 |
| Scotch (Heywood) cannel . . . . . . | 16·8 | 53·94 | — | — | — | 46·06 |
| Scotch (Lesmahagow) cannel . . . . . | } 55·7 | 84·55 | — | — | { $C_3H_8$ } ·91 | 14·54 |
| Whitehill cannel shale | 55·7 | 68·75 | — | 2·67 | — | 28·58 |
| Whitby jet . . . . | 30·2 | 10·93 | — | — | 86·90 | 2·17 |

ANALYSIS OF GASES ISSUING FROM THE MOUTH OF THE BESSEMER CONVERTER.

(*Snelus, I. and S. Inst.*)

| | Time after Commencement of Blow. | | | | | | After addition of Spiegeleisen at Bochum Works. |
|---|---|---|---|---|---|---|---|
| | 2 mins. | 4 mins. | 6 mins. | 10 mins. | 12 mins. | 14 mins. | |
| Carbonic acid . . | 10·71 | 8·50 | 8·20 | 3·58 | 2·30 | 1·34 | — | 0·86 |
| Carbonic oxide . . | nil. | 3·95 | 4·52 | 19·59 | 29·30 | 31·11 | 82·6 | 78·55 |
| Oxygen . . . . | 0·92 | — | — | — | — | — | — | 1·32 |
| Hydrogen . . | 88·37 | 0·88 | 2·00 | 2·00 | 2·16 | 2·00 | 2·8 | 2·52 |
| Nitrogen . . . | | 86·58 | 85·28 | 74·83 | 66·24 | 65·55 | 14·3 | 16·38 |

COMPOSITION OF GASES IN BLOWHOLES OF STEEL INGOTS.
(*Stead, Clev. Inst. Eng*)

| | Steel containing C = 0·42 p. c. Si = 1·00 p. c. mn = 1·08 p. c. | Steel containing C = 0·33 p. c. Si = 0·10 p. c. mn. = 0·69 p. c. | Steel containing C = 0·17 p. c. Si = 0·09 p. c. mn. = 0·89 p. c. |
|---|---|---|---|
| Hydrogen . . . . . | 67·10 | 86·62 | 87·21 |
| Nitrogen . . . . . | 30·30 | 13·29 | 11·15 |
| Carbonic oxide . . . | 2·60 | 0·32 | 1·64 |
| Oxygen . . . . . . | — | 0·37 | — |

AVERAGE COMPOSITION OF COAL GAS.

(*Thorpe.*)

| | |
|---|---|
| Hydrogen . | 45·58 |
| Methane . | 34·90 |
| Carbon Monoxide . | 6·64 |
| Ethene . | 4·08 |
| Quartene . | 2·38 |
| Sulphuretted hydrogen . | 0·29 |
| Nitrogen . | 2·46 |
| Carbonic acid . | 3·67 |
| | 100·00 |

COMPOSITION OF VARIOUS GASES.
(By Weight.)

| | N | H | $CO_2$ | CO | $CH_4$ | $C_2H_4$ | Authority. |
|---|---|---|---|---|---|---|---|
| Blast furnace, Scotch . | 48·20 | 0·90 | 21·70 | 29·24 | — | — | Bell |
| Blast furnace, Askam . | 52·59 | 0·14 | 13·47 | 33·80 | — | — | Crossley |
| Blast furnace, Cleveland | 58·54 | 0·06 | 14·32 | 27·03 | — | — | Stead |
| Producer gas, Siemens . | 64·50 | — | 6·95 | 24·92 | 0·89 | 2·73 | Snelus |
| Producer gas, Siemens . | 63·22 | 0·65 | 8·71 | 25·97 | 1·45 | — | ,, |
| Producer gas, Wilson . | 61·70 | 0·90 | 6·91 | 29·58 | 0·91 | — | Stead |
| Producer gas, Wilson . | 62·84 | 1·11 | 8·29 | 26·33 | 1·43 | — | ,, |
| Retort-made town gas . | — | 8·17 | — | 18·61 | 56·10 | 17·12 | — |

## Oil Gas.

The following are interesting results obtained by Dr. Macadam on experimenting with Pintsch's and Keith's apparatus.

BLUE PARAFFIN OIL USED.

| | Pintsch's Apparatus. | | | Keith's Apparatus. | | |
|---|---|---|---|---|---|---|
| | A | B | Average. | A | B | Average. |
| Specific gravity of oil . | 877·6 | 878·2 | 877·9 | 874·1 | 877·6 | 875·9 |
| Flashing pt. . . . . | 296° | 294° | 295° | 292° | 286° | 289° |
| Firing pt. . . . . . | 356° | 352° | 354° | 348° | 346° | 347° |
| Gas per gallon, cubic feet . . . . . | 90·7 | 103·4 | 97 | 85 | 84·8 | 84·9 |
| Illuminating power . . | 62·5 | 59·1 | 60·8 | 63·2 | 59·5 | 61·4 |
| Volume of oil flowing into each retort per hour . . . . . | 1·4 | 1·18 | 1·29 | 2·3 | 1·3 | 1·8 |
| Gas per retort per hour . | 126·8 | 122·5 | 124·6 | 197·5 | 111·9 | 154·7 |
| Heavy hydrocarbons per cent. . . . . . | 39·2 | 37·1 | 38·2 | 39·9 | 38·2 | 39·0 |
| Gas per ton (cubic feet) . | 23,128 | 26,356 | 24,742 | 21,772 | 21,671 | 21,721 |

PRINTED BY J. S. VIRTUE AND CO., LIMITED, CITY ROAD, LONDON.